Scroll Saw Holiday Puzzles

By June and Tony Burns

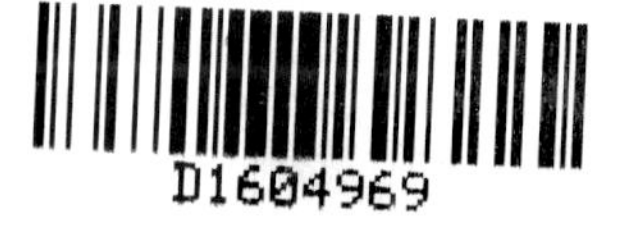

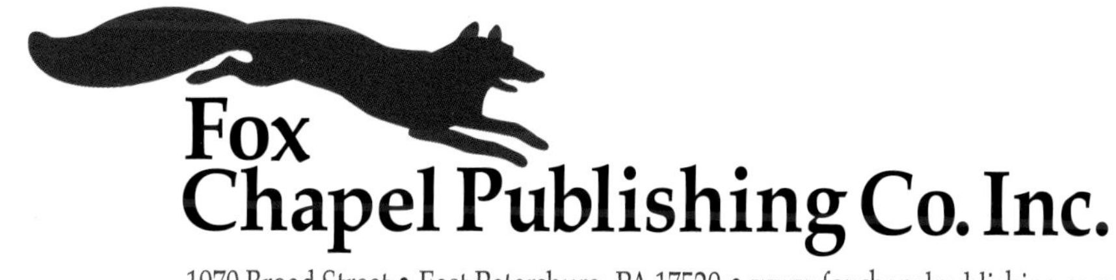

1970 Broad Street • East Petersburg, PA 17520 • www.foxchapelpublishing.com

Dedication
We dedicate this book to our families, who make the holidays very special for us.

Acknowledgments
We wish to thank our children, Jeremy, Emily, Anne and Holly, for their inspiration and our parents for their continued support.

Publisher	Alan Giagnocavo
Book Editor	Ayleen Stellhorn
Step-by-Step Photography	Dr. Thomas F. Burns
Cover Design	Tim Mize
Desktop Specialist	Alan Davis

ISBN 1–56523–204–6

To order your copy of this book,
please send check or money order
for the cover price plus $3.50 shipping to:
Fox Chapel Publishing Co.
Book Orders
1970 Broad St.
East Petersburg, PA 17520

Or visit us on the web at www.foxchapelpublishing.com

Printed in China

Table of Contents

About the Authors

Tony and June Burns

Tony and June welcome your comments and questions, as well as your suggestions for future works. Please include a self-addressed, stamped envelope for a response.

Tony and June Burns
4744 Berry Road
Fredonia, New York 14063

You can also visit Tony and June on the web at **www.scrollsawpuzzles.com**.

Antique scroll saws from the authors' collection.

Introduction

Holidays are a part of every family, every culture. From the new year to the sacred winter holidays, we gather with family and friends each year to share these universal joys. Traditions hold our families and cultures together.

We all know how the holidays make us feel when the snow starts to fall and we hear the first music of the Christmas season. We get a special warmth inside that just cannot be described. The kids can't stop thinking about Christmas vacation, Santa coming down the chimney, and the toys that Grandpa and Grandma will bring when they come to visit.

Our holidays and beliefs help to make us who we are and are an integral part of our lives. Some of our most cherished possessions are those given to us by our parents to commemorate a holiday—whether it is a crèche for Christmas or a menorah for Hanukkah. In our home, we proudly display a crèche that Tony's mom bought at a Grant's Department Store for a few dollars when he was a small child. Perhaps a few projects from this book will be in your family long after you are gone, bringing warm memories to your children and grandchildren.We hope the designs in this book will bring the joys of every season to your home. Celebrate all year while creating some new traditions to share with those you love.

Enjoy,
Tony and June Burns

Getting Started

Selecting wood

Learning about the density of each particular type of wood is important. This will help you to determine how strong a piece of wood is for a particular purpose. It will also help you to assess how easily the wood will cut and what type of blade to use.

We prefer to use silver maple, basswood and pine. These woods are readily available in our area and are relatively easy for beginners to cut. Denser woods and plywood are stronger but harder to cut. We have had fairly good success with 19mm solid core Baltic or Finnish birch.

We also have a great love for oak and butternut, the latter being easier to cut. The harder woods, including Baltic and Finnish birch, tend to dull blades faster, but add fun and challenge to a project. Your local supplier may have these or other varieties from which to choose.

When purchasing wood for puzzles, we suggest that you buy the best grade that you can afford. Look for wood with a straight grain and few defects. Also avoid lumber which appears damp or heavier than usual. The following are the most common defects in wood that affect puzzle making.

• *Warping*: Avoid any wood that is obviously warped or twisted.

• *Cracks*: Cracks occur in the end grain and along the grain of many boards. They are hard to see. A bright light will help to make them visible.

If a board has cracks (known as checking) along the grain, it is worthless. Even if you think you can squeeze your pattern on after the crack, there is most likely another that goes unnoticed. Any wood from the same lot may have the same checking effect, since it was probably cut from the same tree or improperly dried.

• *Knots*: You can work around most knots, but many times they are visible on only one side of the board. Check both sides before laying out a pattern.

Preparing to cut

Lubricate the table: We have found that lubricating the table with paste wax helps to reduce friction and allows your project to move freely, especially when your saw table is cast iron. The lubricant helps to keep the table from rusting and inhibiting your project's movement. (Heavily coat the table if it's going to sit for an extended amount of time between uses.)

Square the blade: It is a good idea to occasionally check the "squareness" of your cut. This can be done on a weekly basis or as needed. To do so, use a small 2-inch engineer's square. Adjust the table to 90 degrees. Check the blade thickness: A micrometer is another good investment. Being able to verify the thickness and the width of a blade helps to troubleshoot problems.

Covering the scroll saw's table with Plexiglas™ will help to reduce friction, making it easier to cut the pieces. The cover is cut from 1/4 in. Plexiglas™ and attached to the table with magnetic strips. A hole is drilled through the center to correspond with the path of the blade.

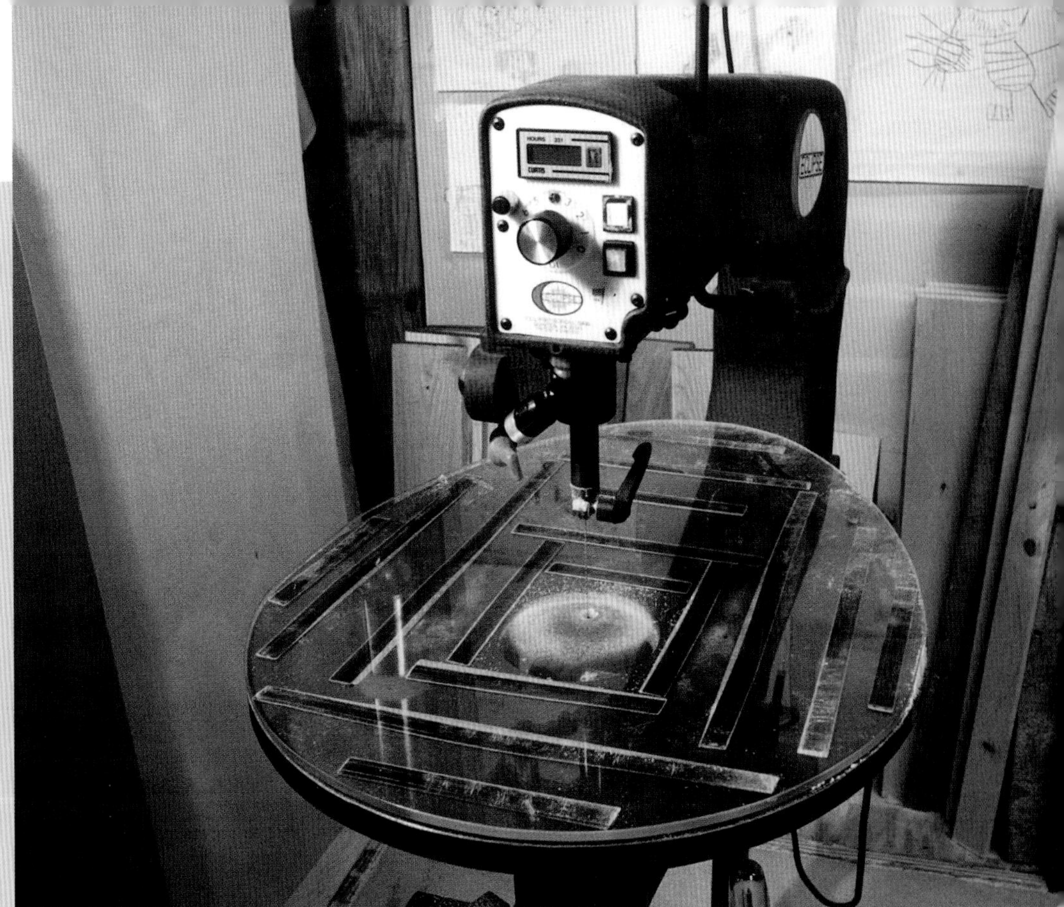

It is also a good way to verify a blade size in the event the blades get mixed up.

Light the area: Good lighting is necessary when cutting. Incandescent light is easier on the eyes than fluorescent. If your saw is not sold with a light, most manufacturers sell a light kit or a magnifying lens separately.

Buy an anti-vibration mat: Placing one under the saw reduces vibration and operator fatigue.

Use safety gear: Three areas of safety concern are ears, eyes and respiration. A quality set of comfortable hearing protectors or ear plugs are a must if you cut for any extended period of time. For proper eye protection, quality safety glasses with side shields are necessary. If you wear tempered prescription glasses you can purchase eye shields for them.

As avid scrollers, we are concerned with the quality of the air we breathe when we operate machinery. A good quality respirator is necessary to trap the fine dust and particles in the air. In addition to this, we have both a permanent air filtration system mounted on the ceiling and a portable unit on wheels.

About scroll saw blades

Fret saw (scroll saw) blades are typically made from high carbon steel that is heated, forged and drawn over many miles of rollers. This makes the steel continuously thinner until it becomes wires of the desired thickness. These are then flattened. If you

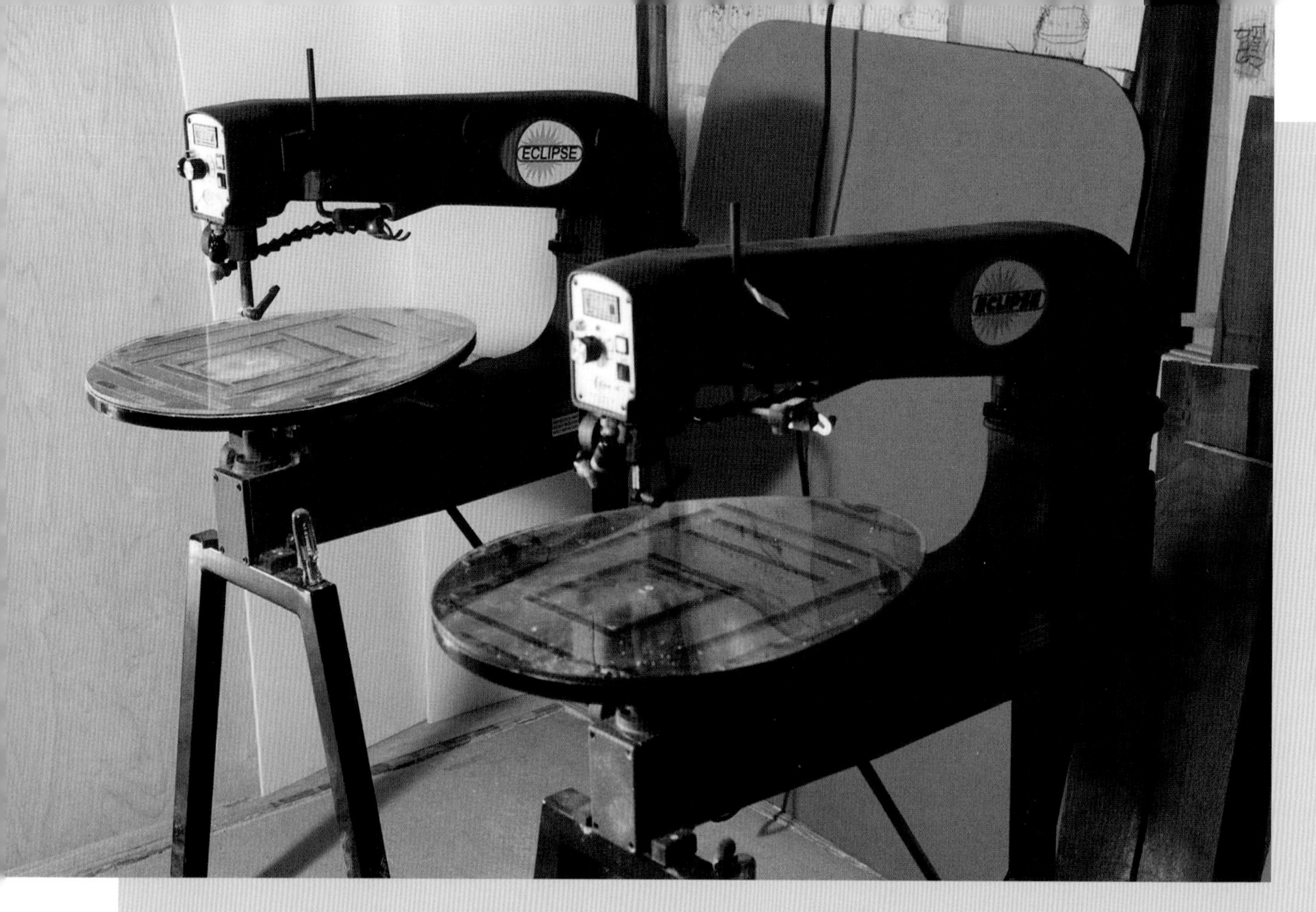

If you own more than one scroll saw, try this set-up. Each of the saws shown here has a different blade. The first saw is threaded with a #7 blade for rough cutting; the second with a #3 blade for fine cutting. Keeping both saws set up saves the time of switching back and forth between the blades.

look closely, you will notice a slight curvature on the back of most blades, an indication that they started out as wires. After the steel is flattened, the particular tooth style and size are ground in .

Blades can also be made by a punch or through a shearing action, but are not typically made this way today; they are not as sharp or uniform as those that are ground. We have a small collection of these handmade older blades that came from treadle machines used during the late 1800s to early 1900s.

Choosing a blade

Blades are the heart and soul of a scroll saw. The finest scroll saw will not work to its full potential if you are using the wrong blade.

For fine cutting 3/4-inch (19mm) or 1-inch wood, we suggest that you start with a #3 or #5 blade. These blades will give you better detail and smoother cuts. Although there is no rule, keep in mind that a smaller blade (such as #0 or 2) will take longer to cut and may burn the wood from extra friction. The larger blades, #9 and 12, are easier to use when you are cutting straight lines or larger radiuses, but are not the best choice for detail. These larger blades do not leave as fine an edge on the wood, requiring more sanding than a finer blade. Larger blades also make the puzzles looser, which may or may not be desirable.

Blade sizes may vary from one manufacturer to another. For example, what one manufacturer calls a #2 blade may be extremely close to another company's #3 blade. TPI (teeth per inch) may also be measured differently from one manufacturer to the next. One company may measure teeth from tip to tip; another may measure from gullet to gullet. This can change the count by half of a tooth to a whole tooth.

It is also important to take care of your blades. When blades are stored for long periods of time they tend to rust. We recommend spraying them with a light coat of oil or WD40 to prevent rusting. We know this trick works first-hand because we use it frequently; we buy all of our blades for the year at the same time.

One of the most common questions we get is: "What is the best blade for my saw?" This question is next to impossible to answer. Each scroll saw handles the same blade differently, so experimenting with different types and sizes of blades is the way to decide which blade works for you. It is one of the best investments you can make.

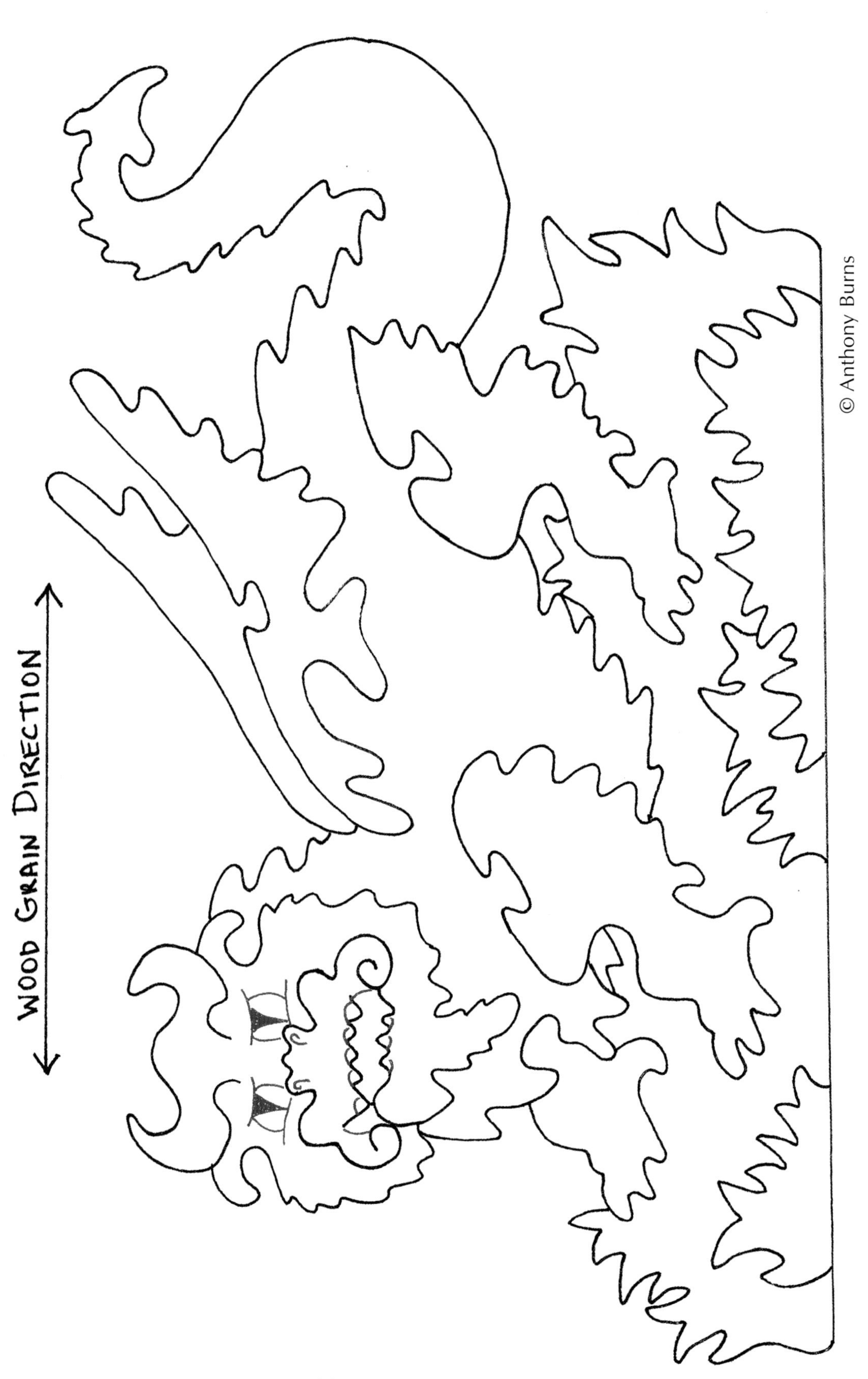

Red line - Detail line
Black line - Cut line

Step-by-Step Chinese Dragon Puzzle

To start this project, you will need a piece of wood slightly larger than the pattern on the facing page. Make sure the grain direction of the pattern matches that of the wood. If you select some form of plywood, such as Baltic birch, the direction of the grain is not important as long as the grain runs up and down or side to side.

How-to

Make a copy of the pattern.

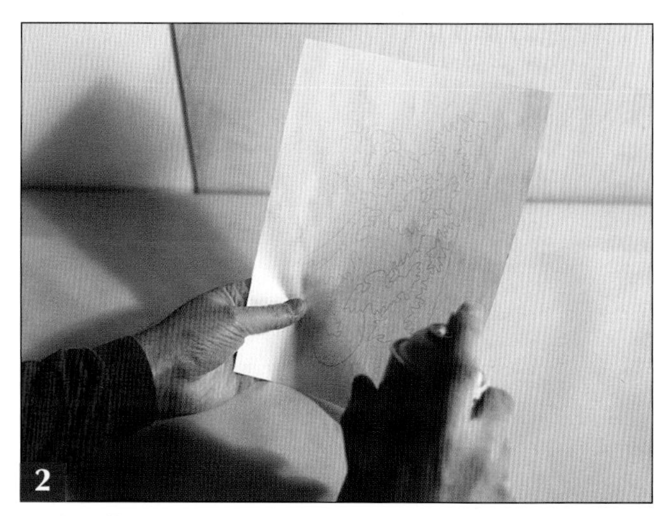

Spray adhesive on the back of the pattern; then trim the pattern and press it onto the wood. (Note: Use spray adhesive in a well-ventilated area.)

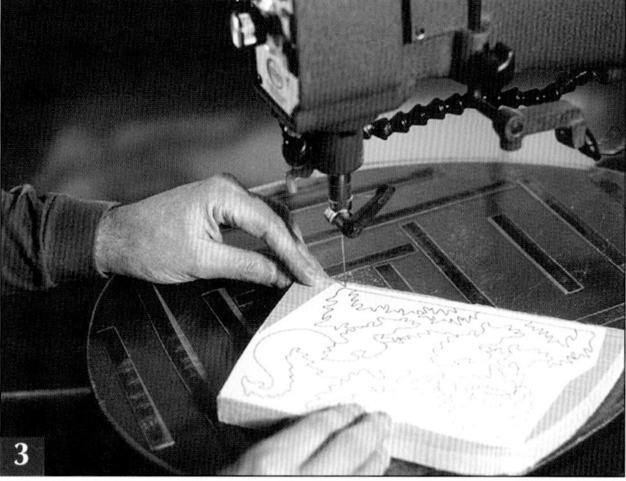

Start cutting at the bottom, near the back of the pattern, using a #7 blade. (Note: The hold-down arm of the saw was moved to the back to photograph the cuts. We recommend you keep the arm in place as you saw.)

Turn the speed down when you make sharp turns. This makes it easier to turn the wood on a sharp angle. The area where the wings meet the body is a good example of an area to proceed at a slower pace.

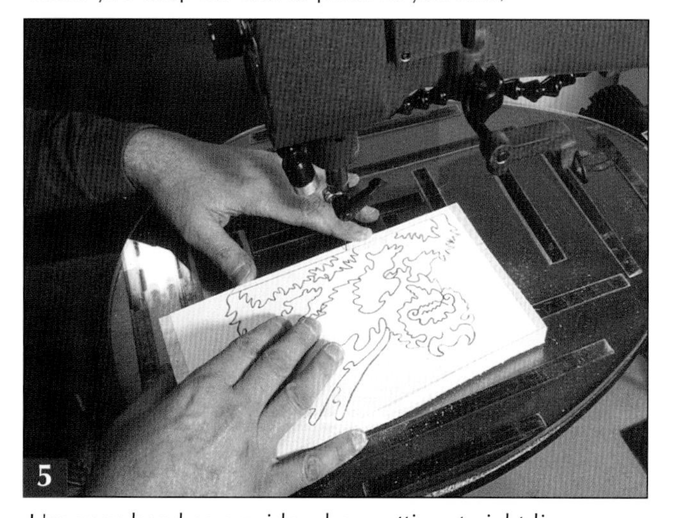

Use your hand as a guide when cutting straight lines.

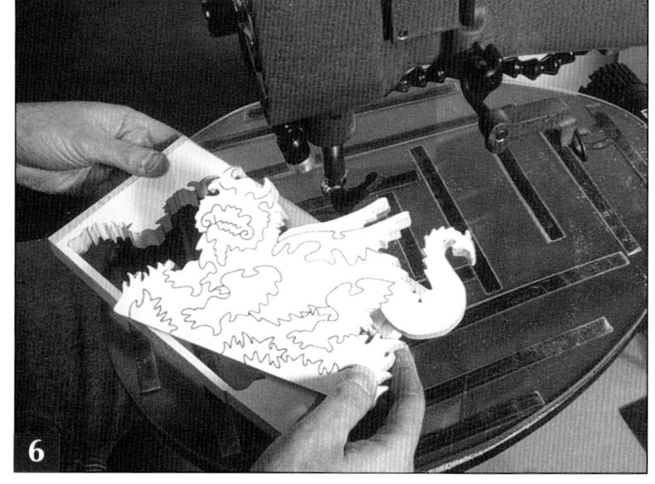

Remove the scrap. This completes the rough cutting of the design.

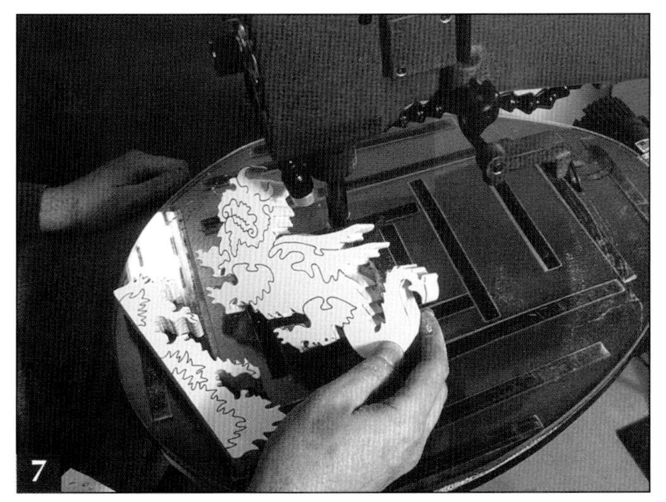

For fine cutting, change to a #3 blade.

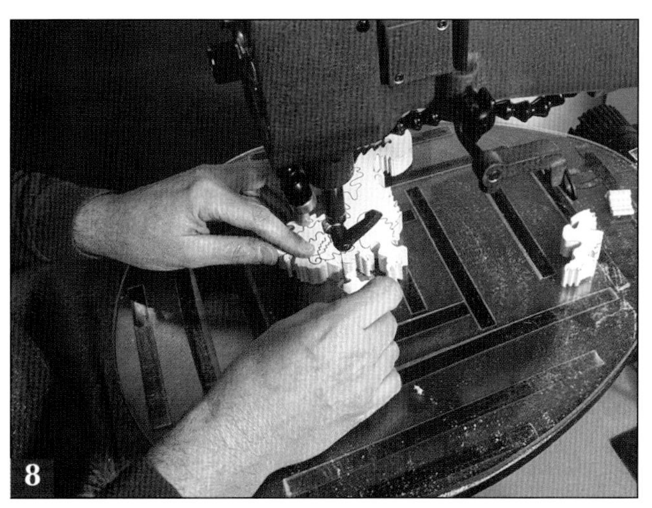

Cut the rear and front legs. Assemble the cut pieces on the back of the saw so you don't lose any.

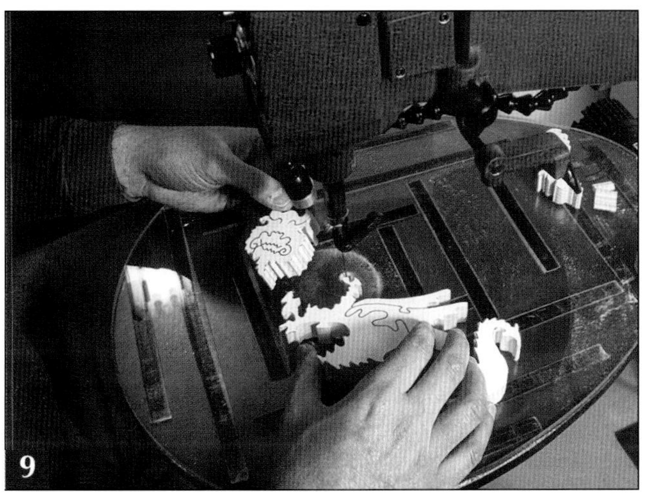

Cut the head.

Cut the details on the horns; then cut the face and the teeth.

How-to

Cut the wings and assemble the dragon.

Using a hairdryer warm the adhesive to remove the pattern. With a sharp object, such as a metal-working scratch awl or the edge of a craft knife, carefully lift the edges of the pattern, using caution to avoid gouging the wood.

Sand the front and back surfaces of the dragon on 150-grit sandpaper adhered to a piece of wood.

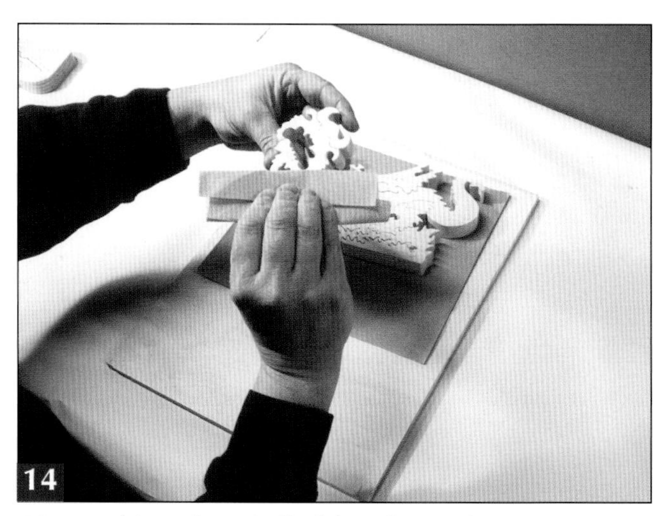

Disassemble and sand all of the edges with 220-grit sandpaper.

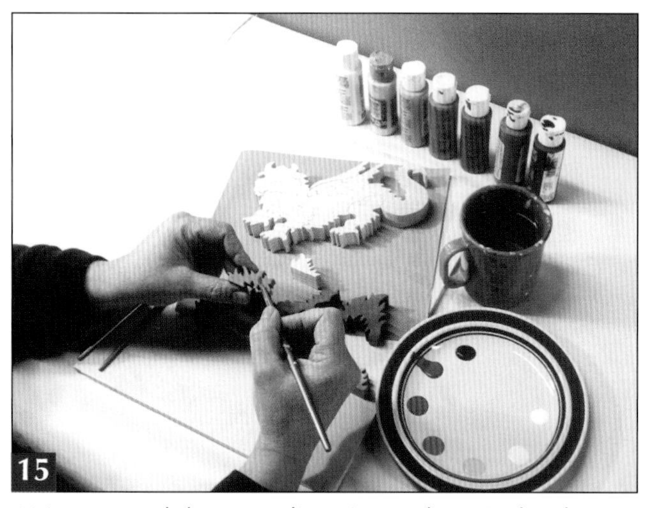

Using watered-down acrylic paints and a paint brush, stain the puzzle pieces. Wipe off any excess paint with a soft rag.

Stain the dragon's body red.

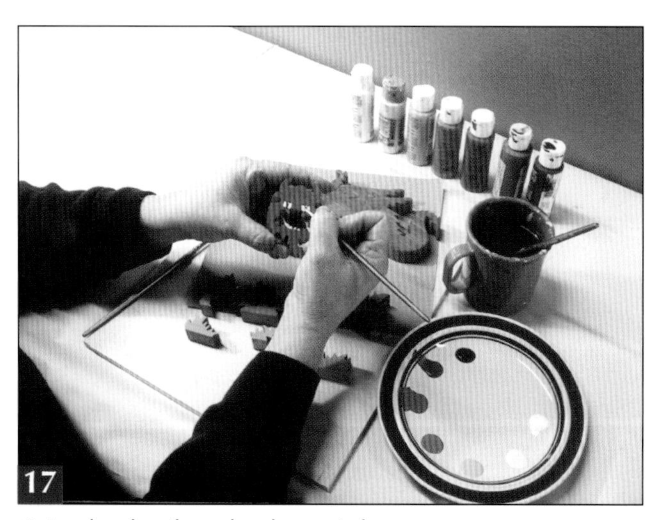

Paint the detail on the dragon's face.

How-to

New Year

Elephants in Champagne

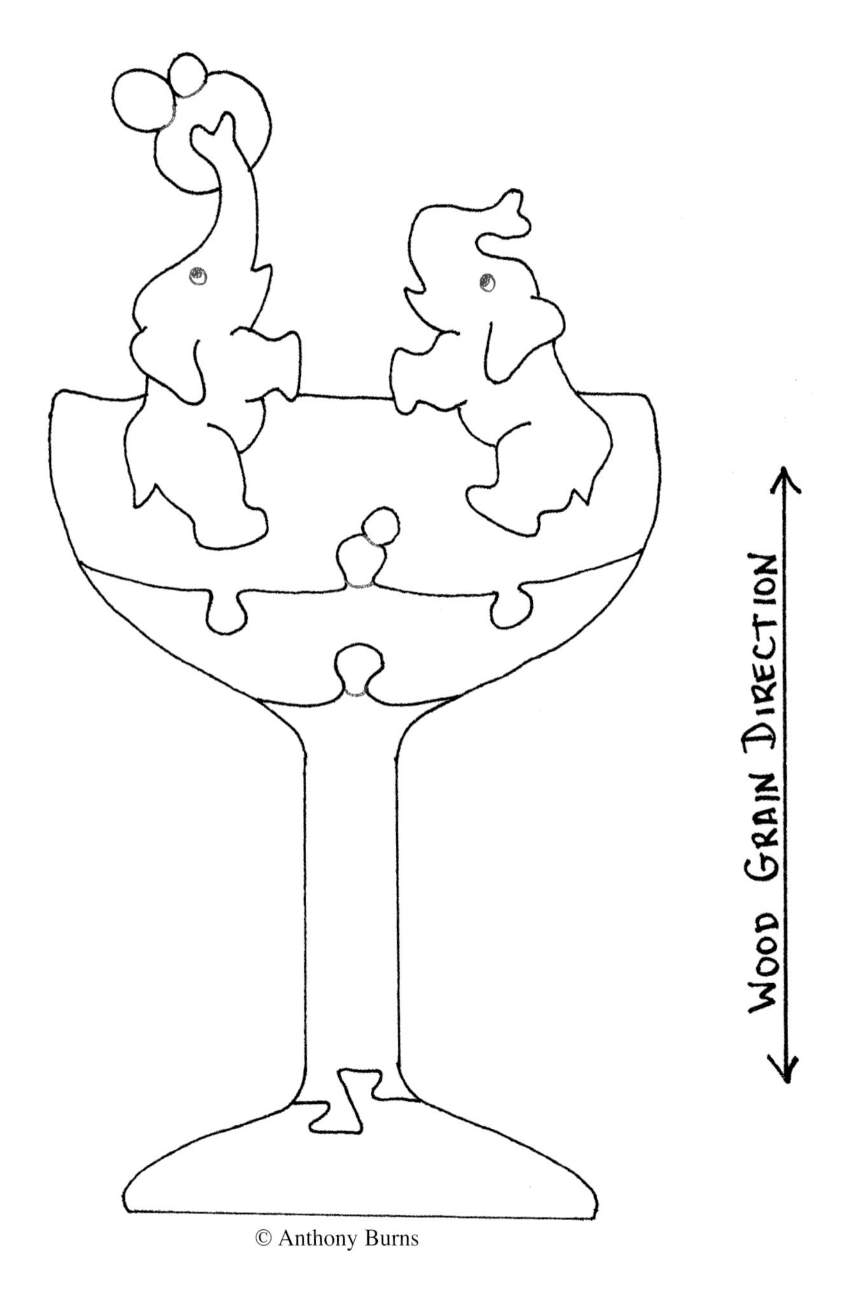

Red line - Detail line
Black line - Cut line

Valentine's Day

Hearts and Doves

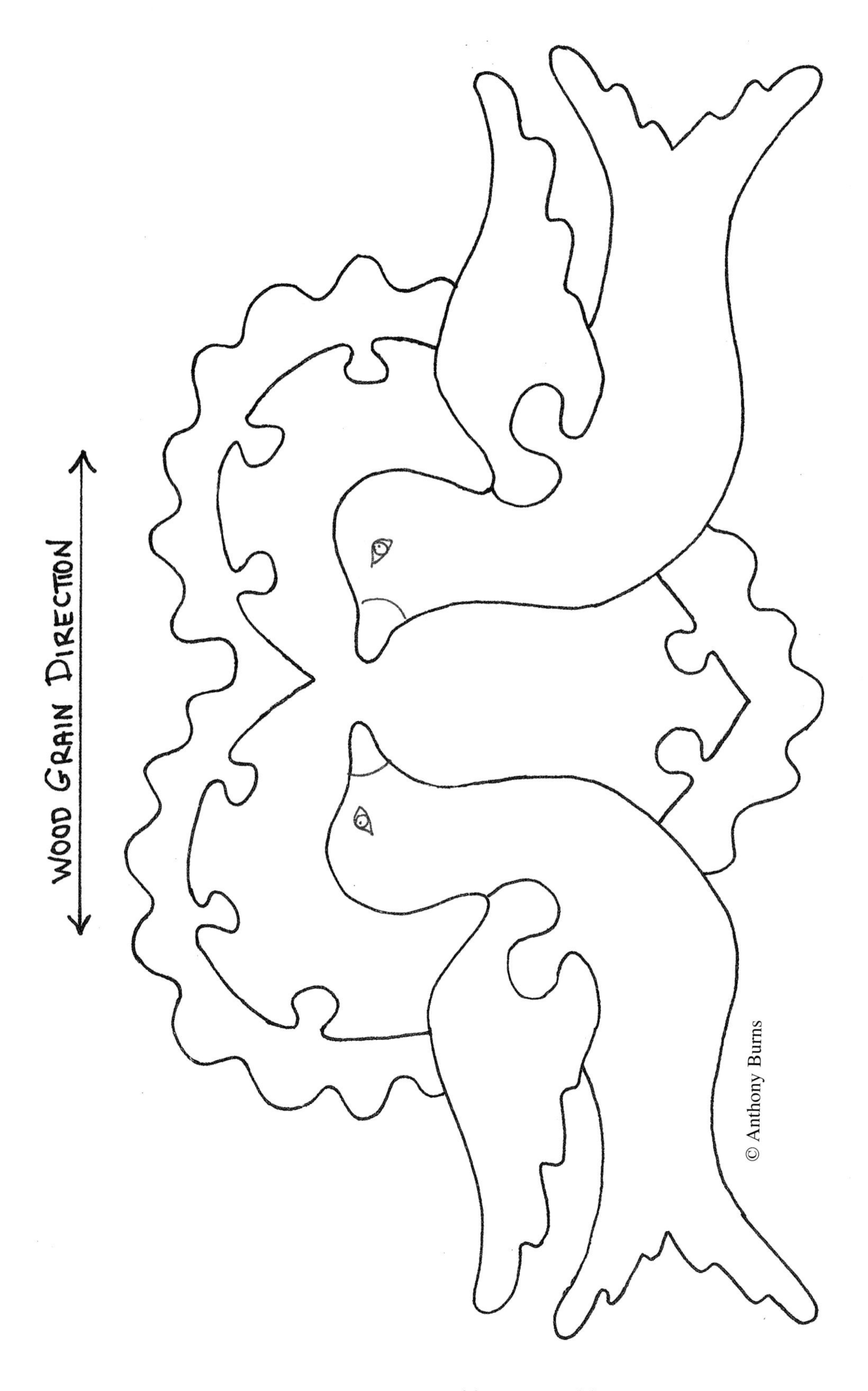

Red line - Detail line
Black line - Cut line

St. Patrick's Day

Leprechaun

WOOD GRAIN DIRECTION

Red line - Detail line
Black line - Cut line

First Day of Spring

Tree and Birds

Red line - Detail line
Black line - Cut line

Easter

Bunnies in a Basket

WOOD GRAIN DIRECTION

Red line - Detail line
Black line - Cut line

Mother's Day

Rose in a Vase

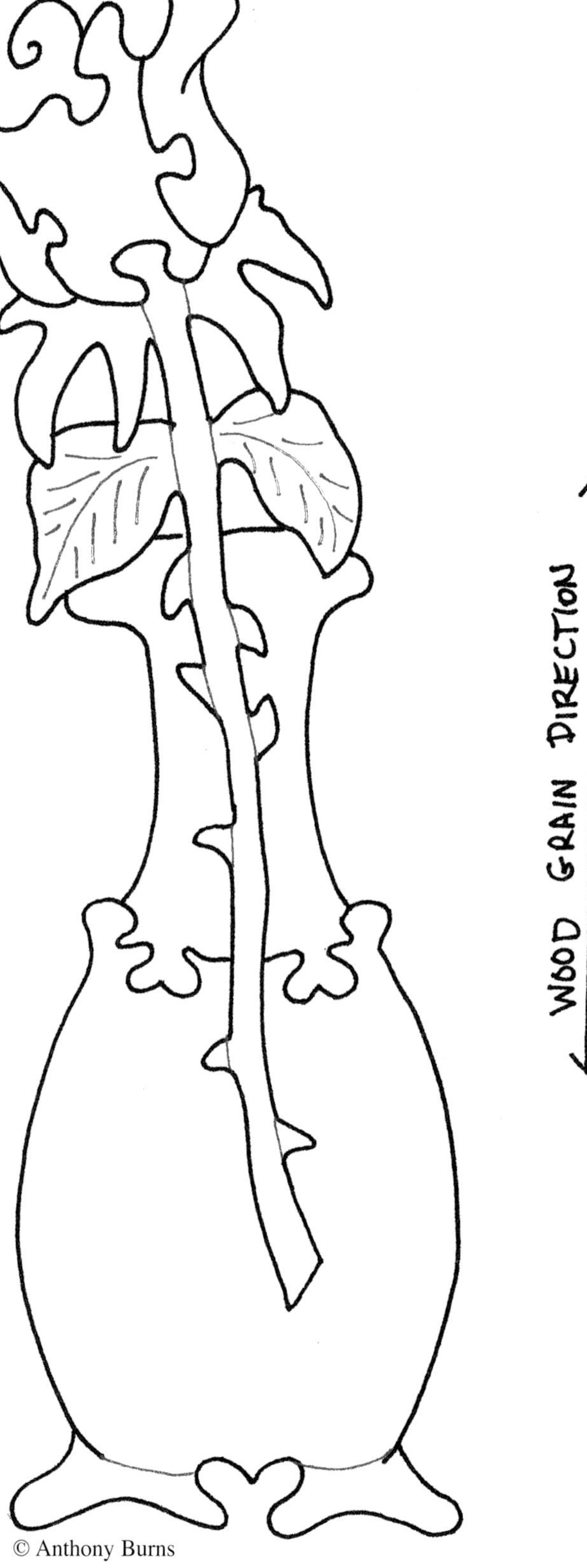

Red line - Detail line
Black line - Cut line

Number One Dad

Red line - Detail line
Black line - Cut line

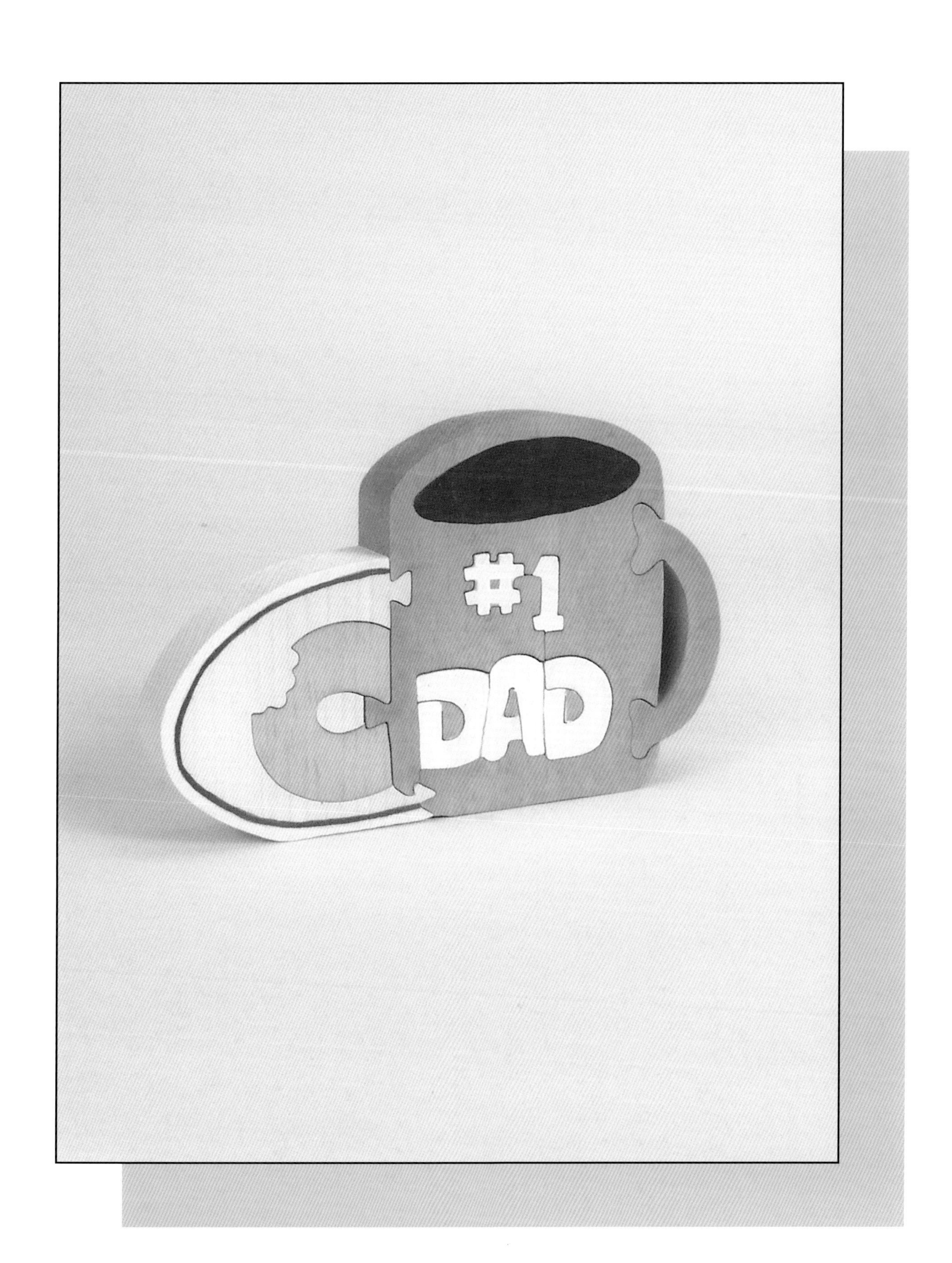
#1
DAD

Fourth of July

Independence Day Parade

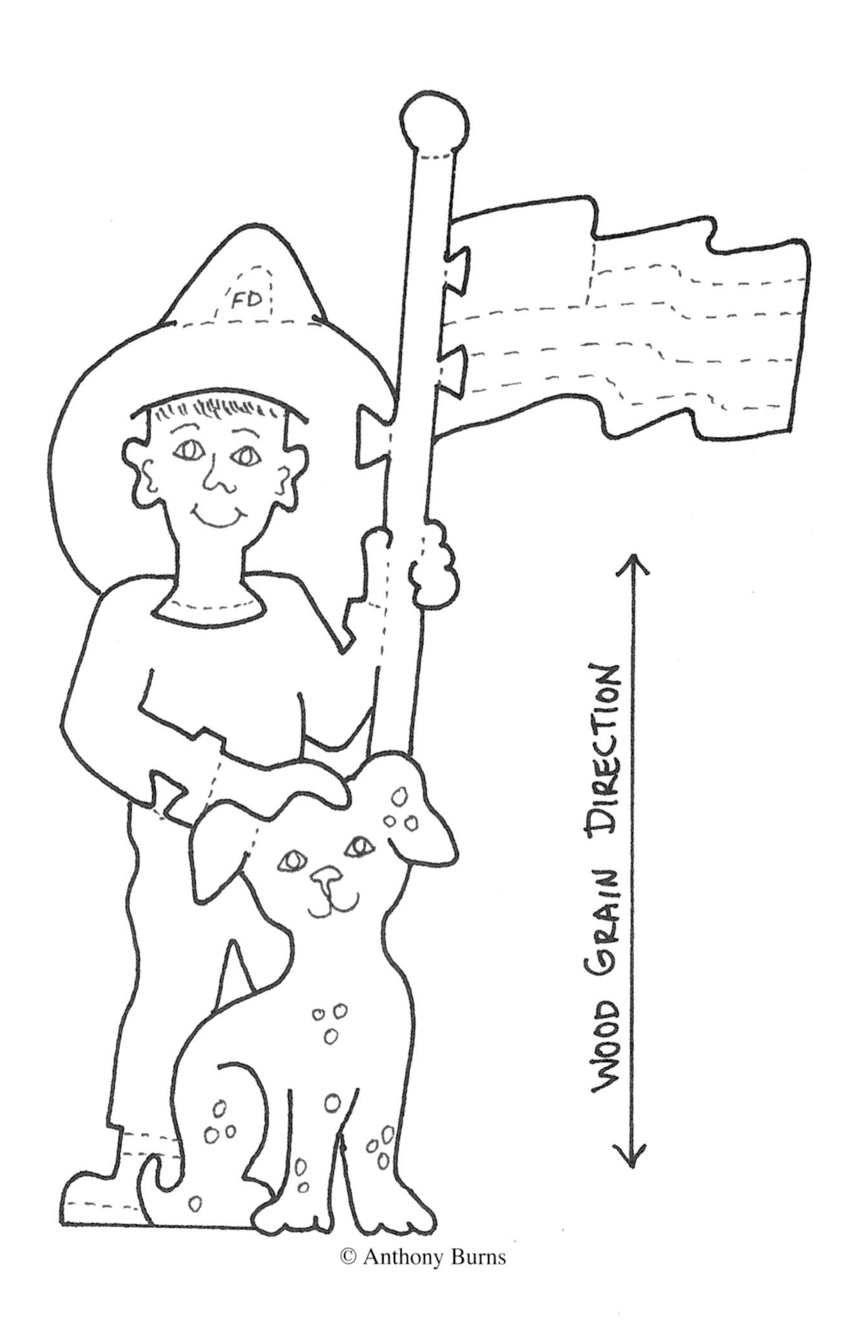

Red line - Detail line
Black line - Cut line

F.D.

Fourth of July

Star Spangled USA

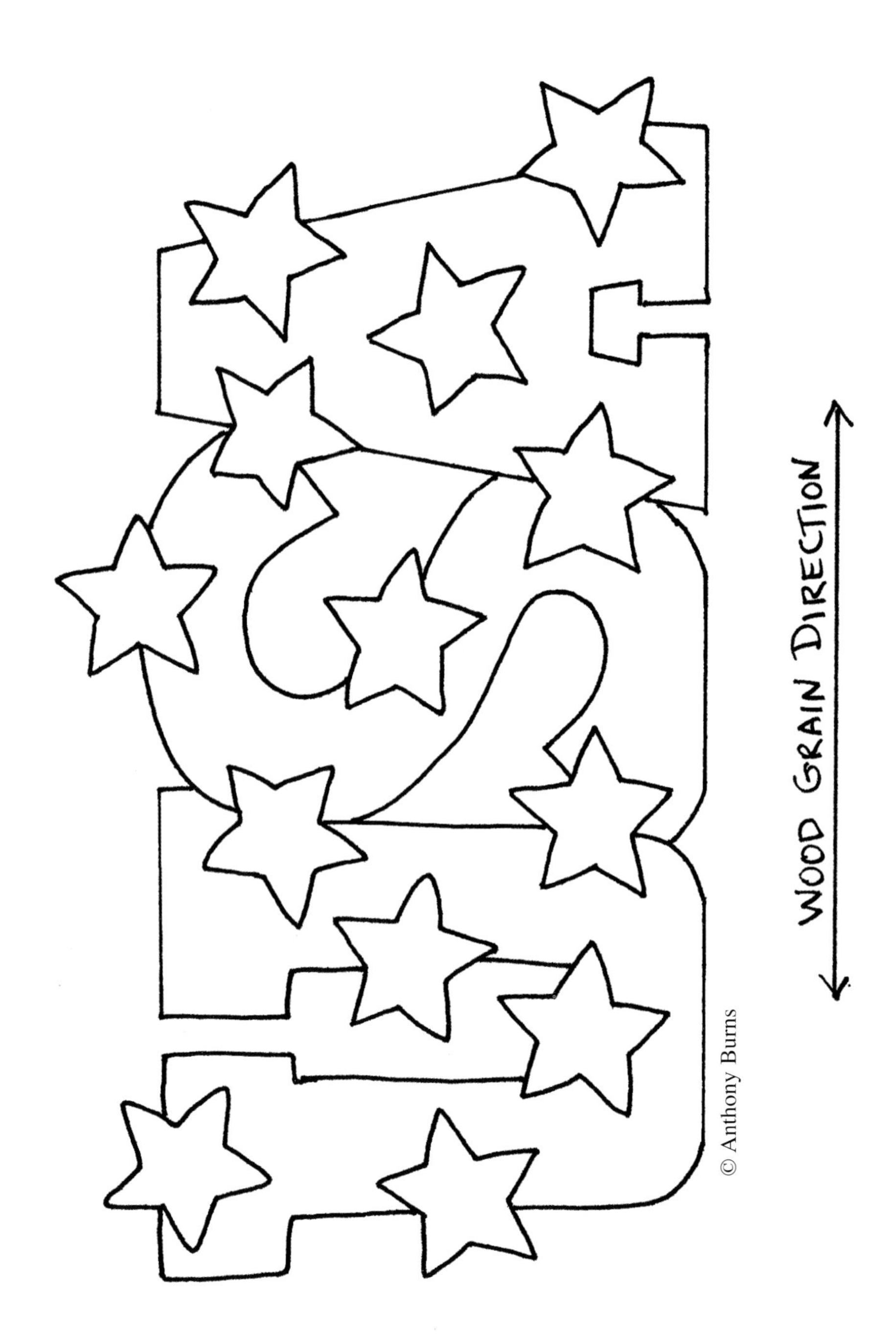

Black line - Cut line

USA

First Day of Summer

Summer Flight

Red line - Detail line
Black line - Cut line

First Day of Autumn

Autumn Wreath

← WOOD GRAIN DIRECTION →

Red line - Detail line
Black line - Cut line

Halloween

Peek-A-Boo Kitty

Red line - Detail line
Black line - Cut line

Halloween

Ghost with Pumpkin

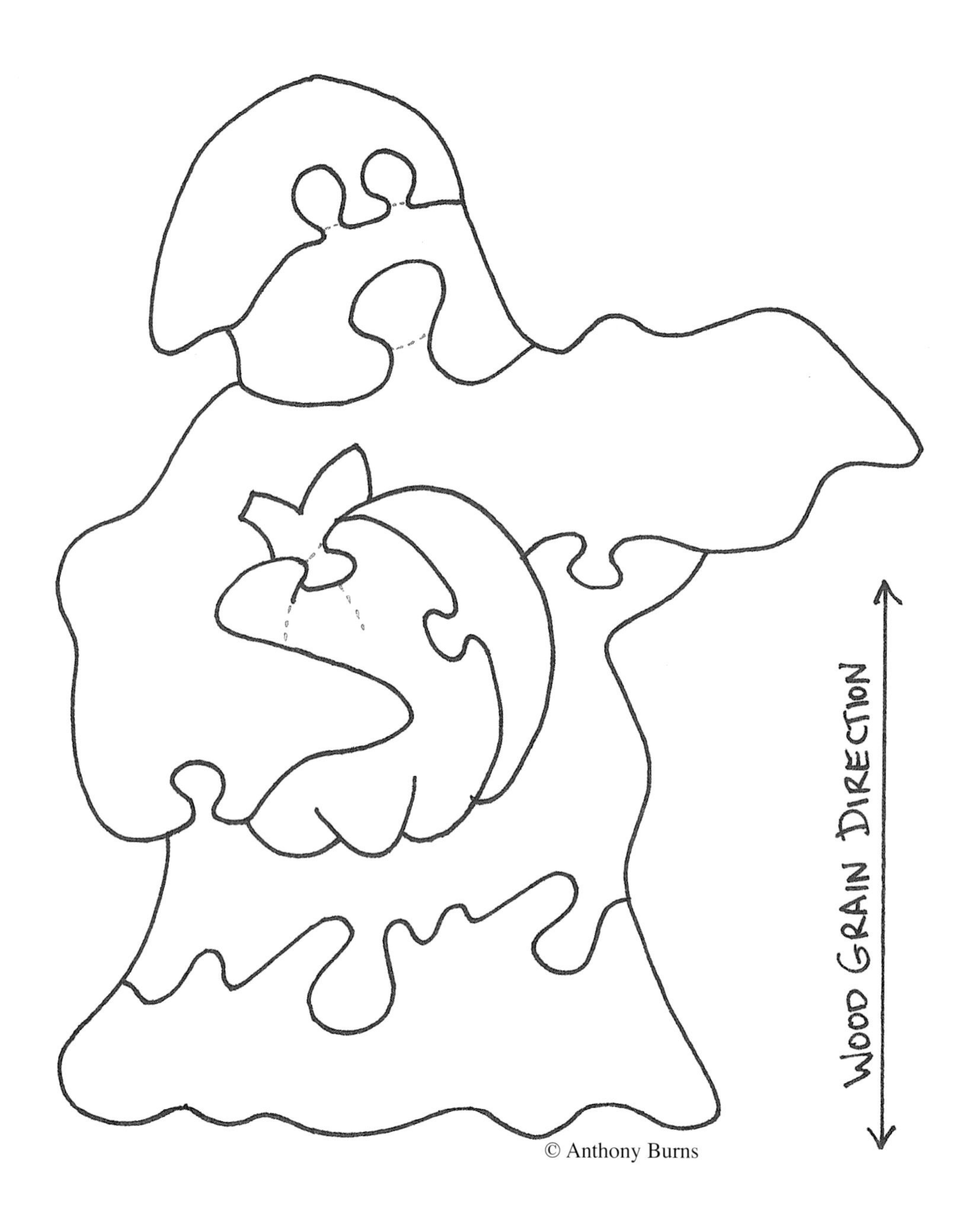

Red line - Detail line
Black line - Cut line

Halloween

Witch in Flight

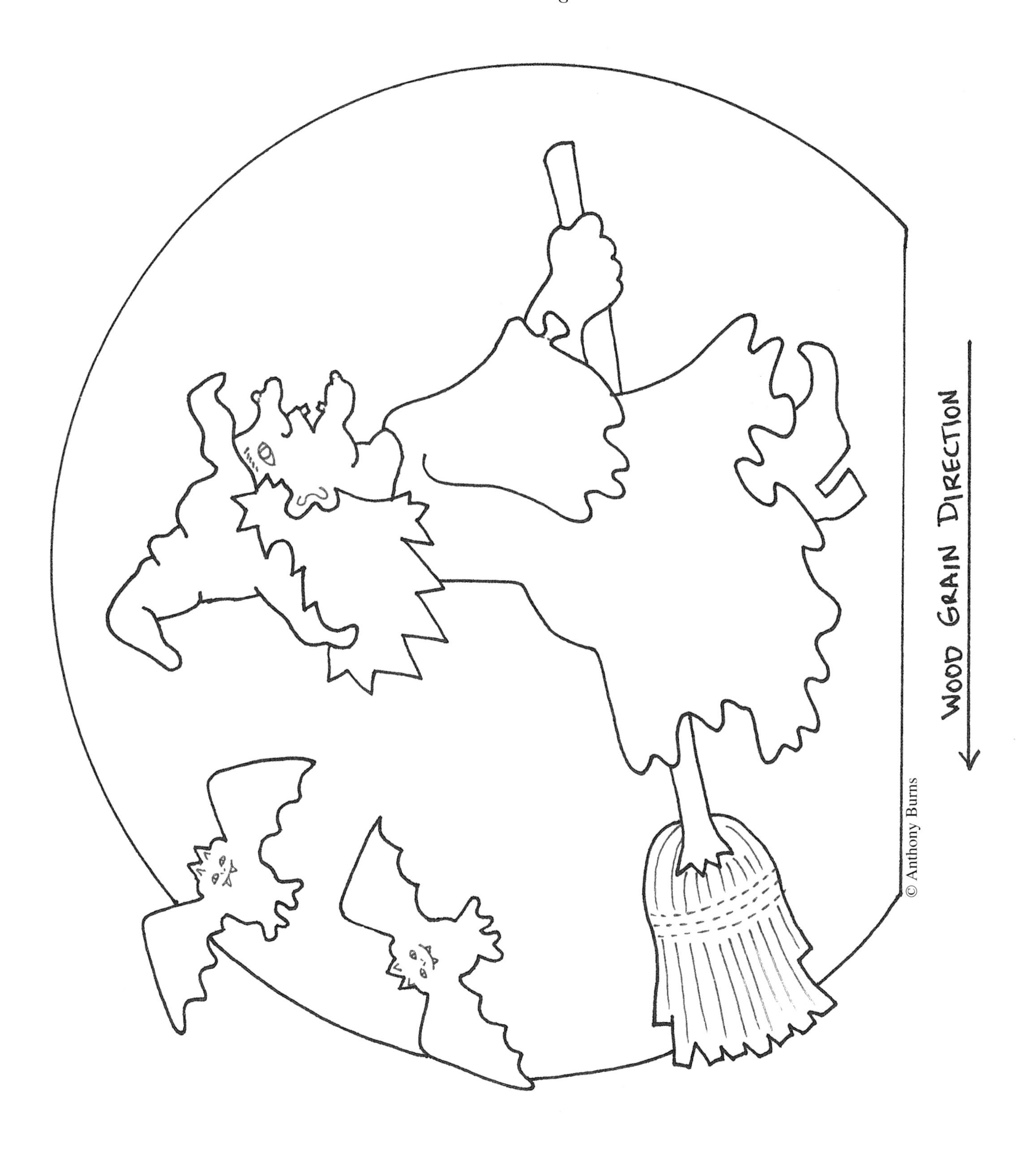

Red line - Detail line
Black line - Cut line

Thanksgiving

Harvest Cornucopia

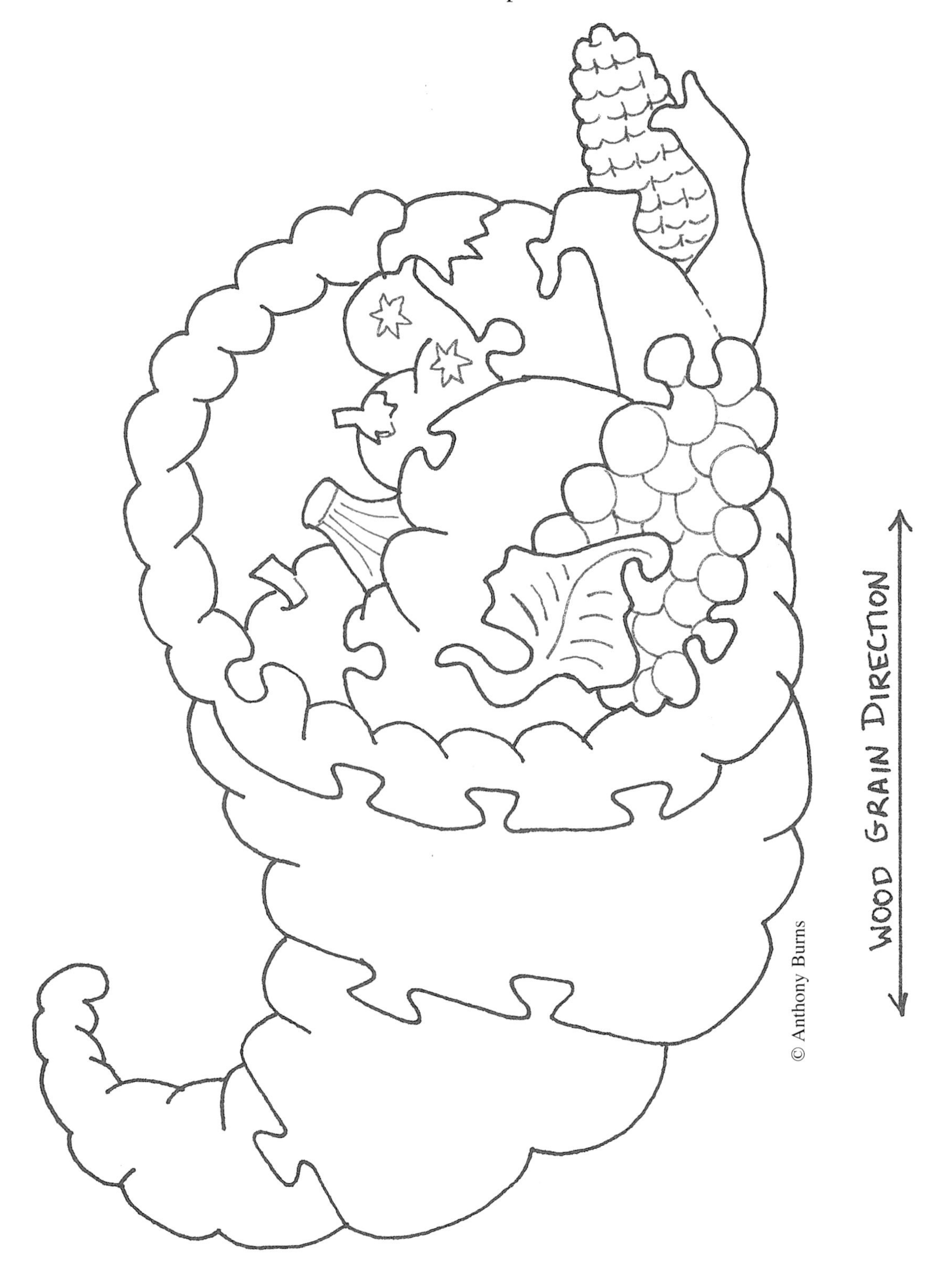

Red line - Detail line
Black line - Cut line

Red line - Detail line
Black line - Cut line

Step-by-Step
Friendly Snowman Puzzle

To start this project, you will need a piece of wood slightly larger than the pattern on the preceding page. Make sure that the grain direction of the pattern matches that of the wood. We selected 3/4 inch pine because it is readily available in our area. Another great choice is poplar. Solid-core Baltic birch also works well, though it tends to wear out blades quickly. You may wish to try out some of your own local varieties of wood or different thicknesses.

How-to

Copy the pattern of the snowman. (Note: I removed the painting detail so it would be easier to see the pattern in the step-by-step instructions.)

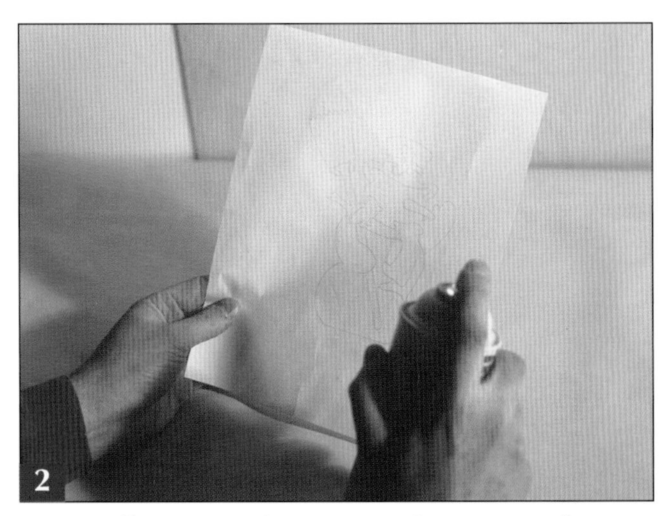

Spray adhesive on the pattern; then trim to fit on the wood.(Note: Use spray adhesive in a well-ventilated area.)

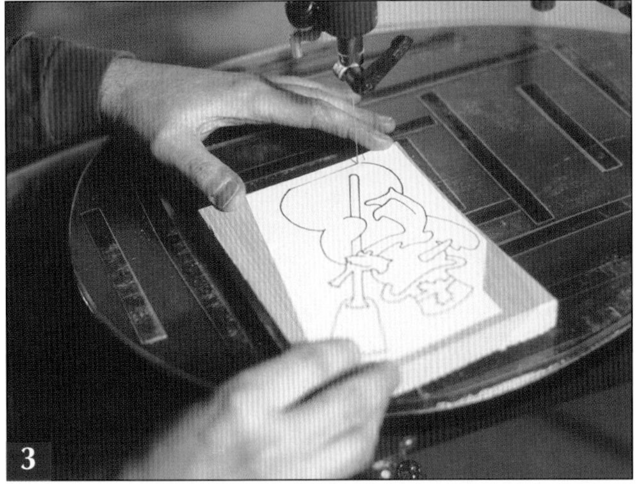

Using a #7 blade, start cutting at the bottom of the pattern so the entry is not noticeable.

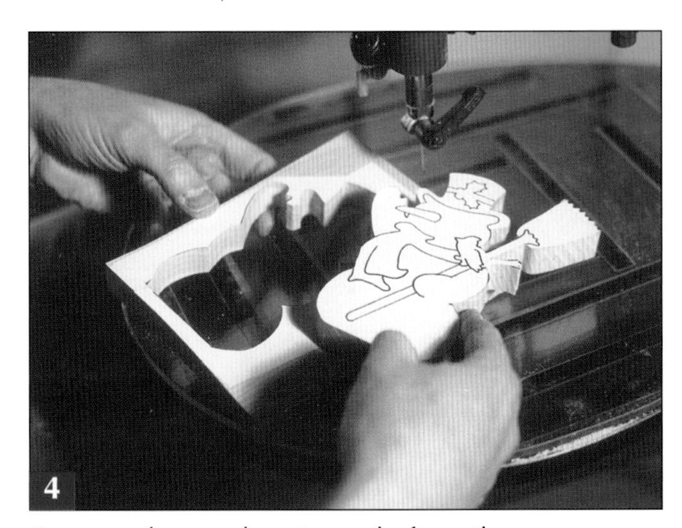

Remove the rough-cut puzzle from the scrap. Discard the scrap wood.

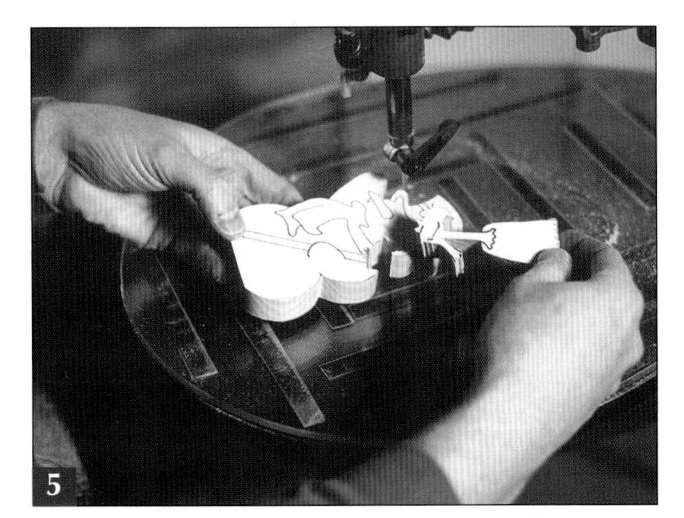

Cut the cardinal and the broom. Remember to cut the detail in the bird before cutting it away from the main pattern.

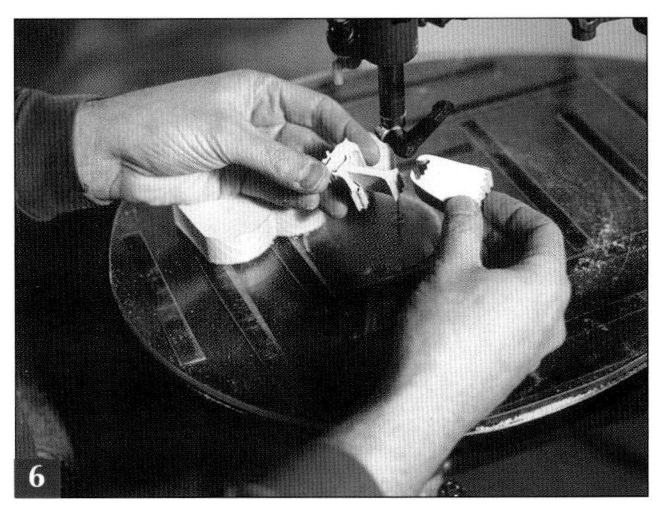

Cut the top of the broom; then reassemble it and put it off to the side.

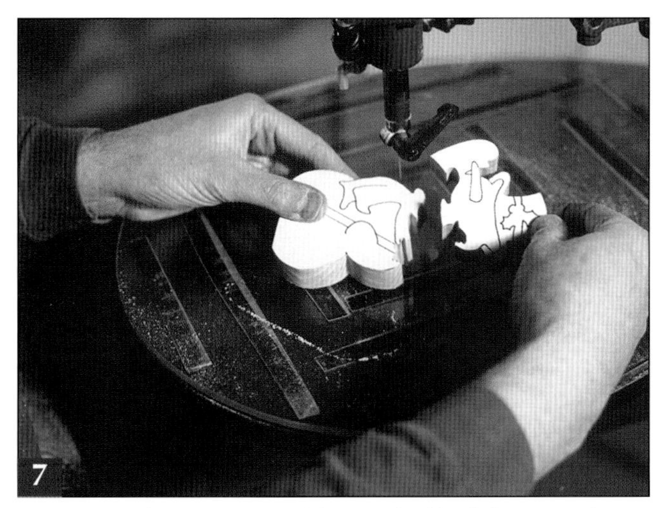

Separate the upper and lower half of the puzzle. This makes it easier to manage, a plus for some of the larger designs.

Cut the holly and remove it from the hat.

Cut the bottom of the hat from the snowman; then cut the carrot nose.

Remove the waste-wood from the bottom of the carrot and the top of the arm.

How-to

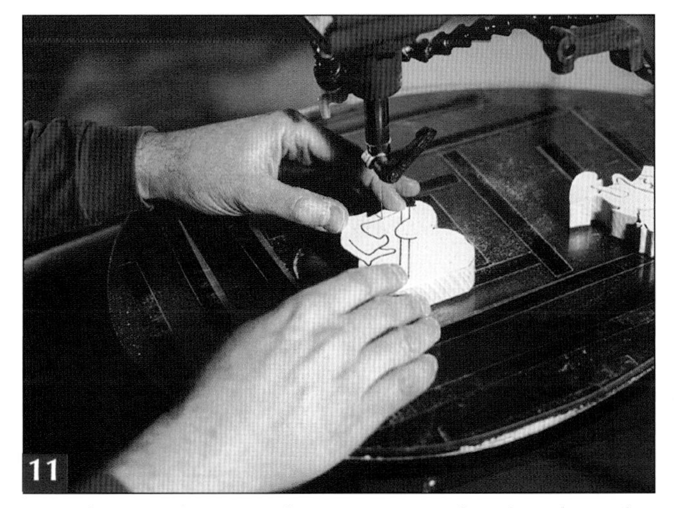

Cut the scarf, remembering to cut the detail on the inside at the same time. It's easier to do those details now, when the piece isn't so small.

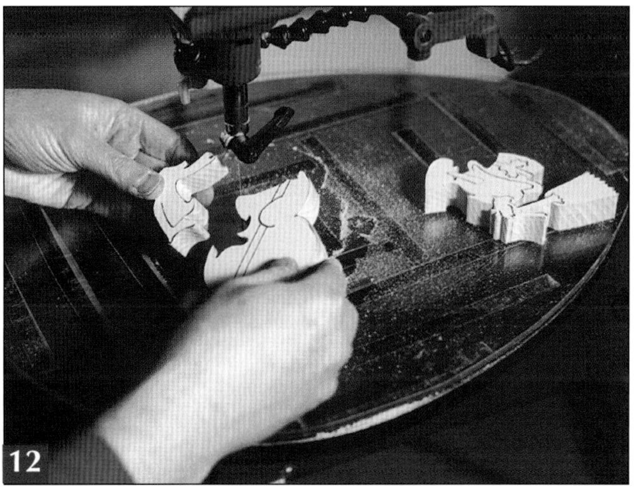

Remove the scarf. You can see the puzzle pieces coming together at the back of the saw!

Cut out the internal broom handle and the detail on the arm.

Assemble the cut pieces into the whole design.

How-to

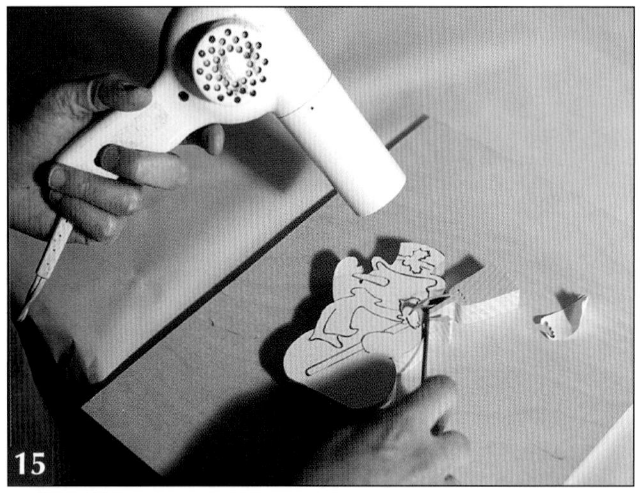

Heat the pattern with a hair dryer and carefully pull it from the wood with a metal scratch awl or similar tool.

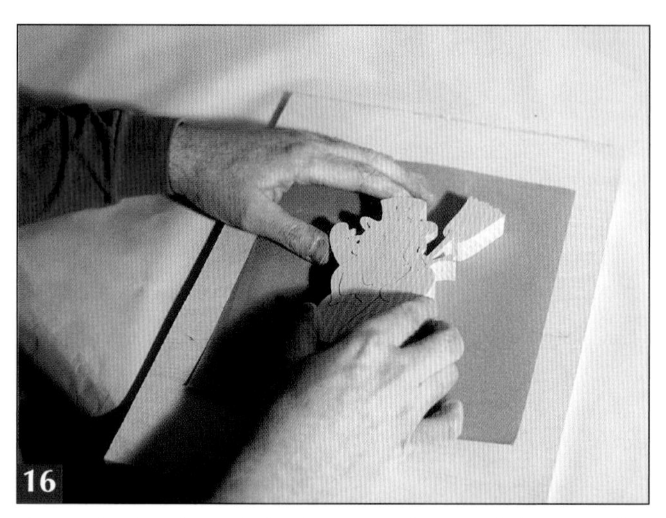

Sand the back and the front surfaces of the puzzle on a sheet of 150-grit sandpaper, moving the puzzle up and down in the direction of the grain.

Disassemble the puzzle and sand each part with 220-grit sandpaper to remove any possible burrs.

Using acrylic paint as a wash, brush the color on the wood; then wipe it off using a soft cloth rag.

Add all details as shown on the pattern; then assemble the finished puzzle.

How-to

Hanukkah

Star of David

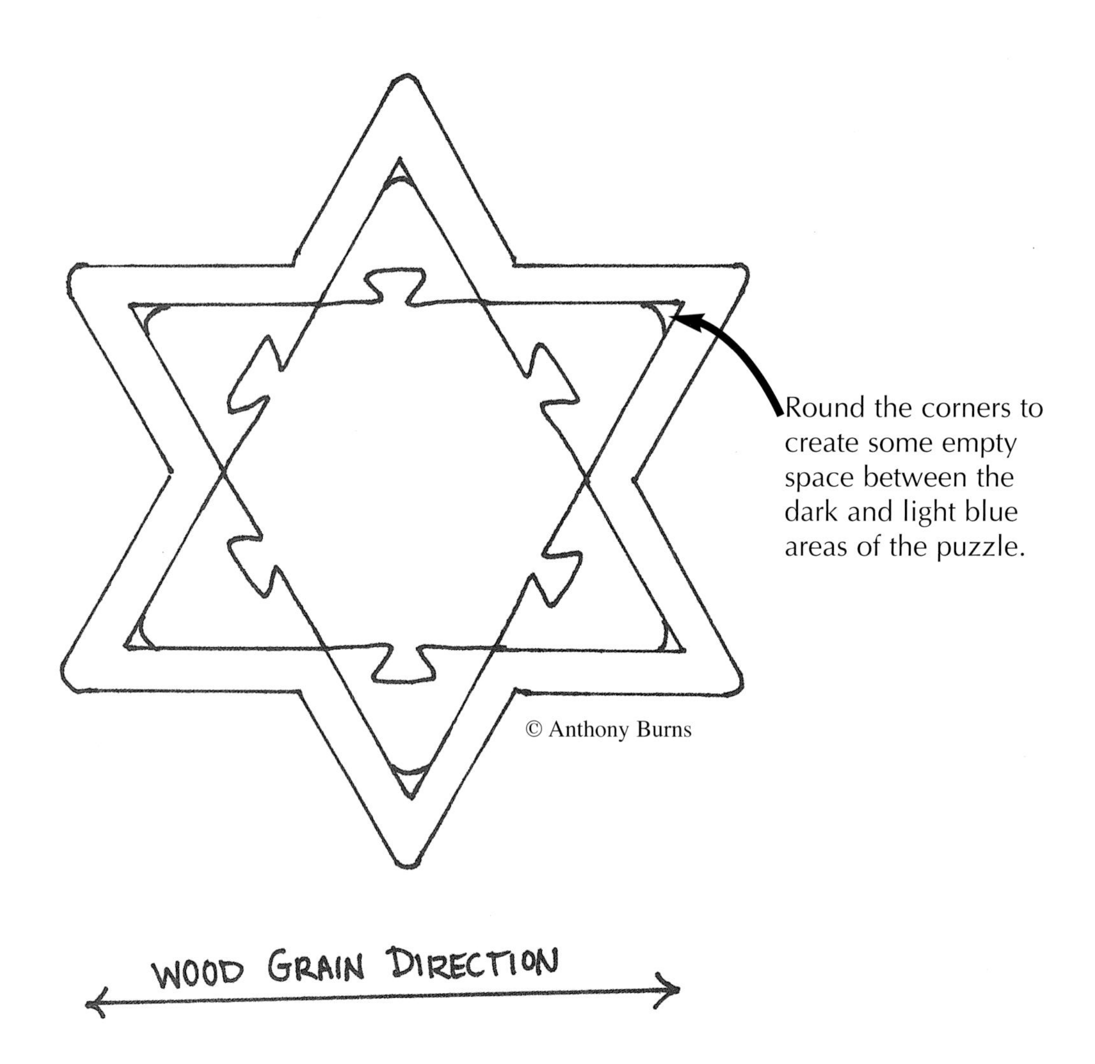
Round the corners to create some empty space between the dark and light blue areas of the puzzle.
© Anthony Burns
WOOD GRAIN DIRECTION

Red line - Detail line
Black line - Cut line

Hanukkah

Menorah

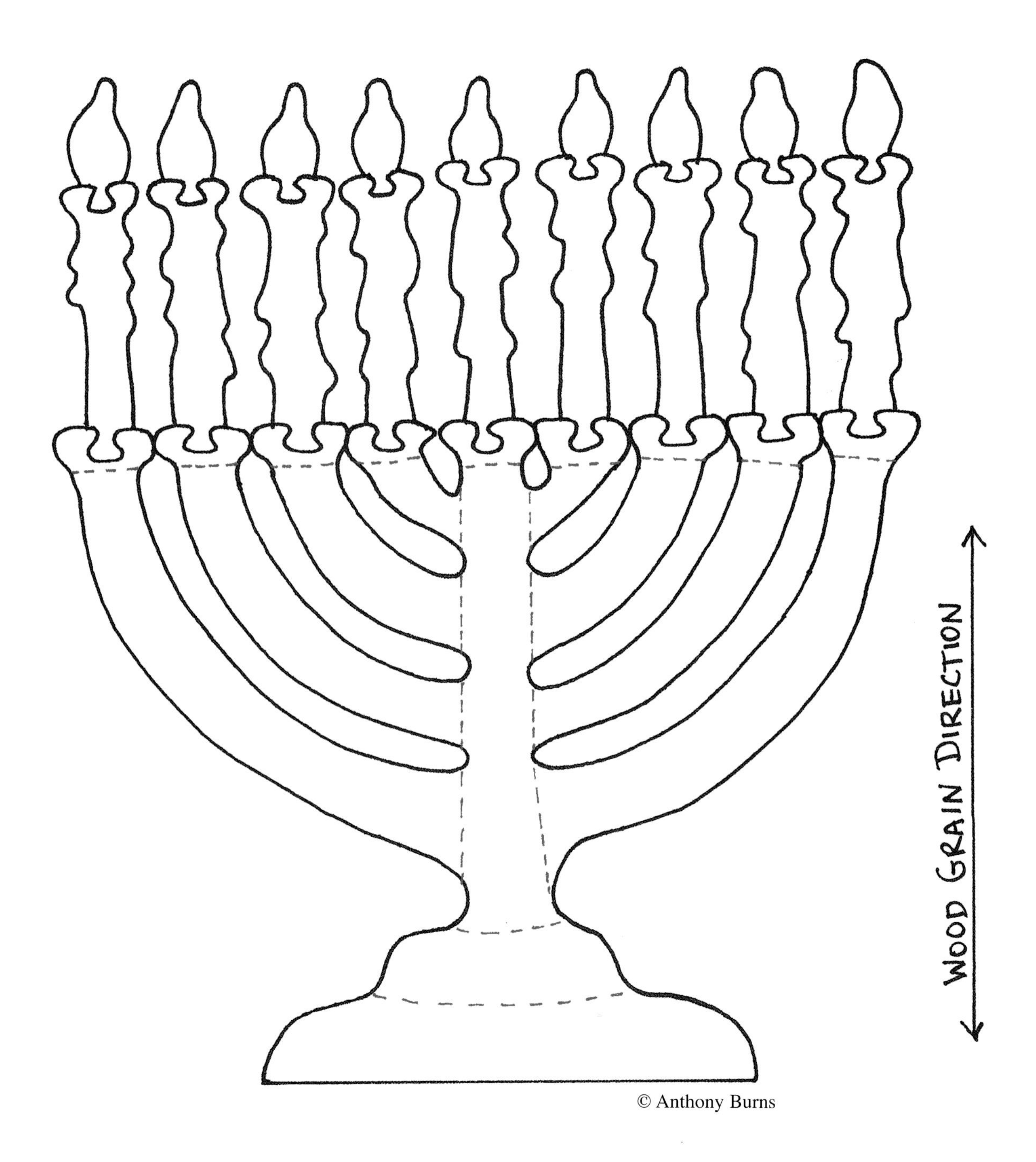

Red line - Detail line
Black line - Cut line

Santa of Peace

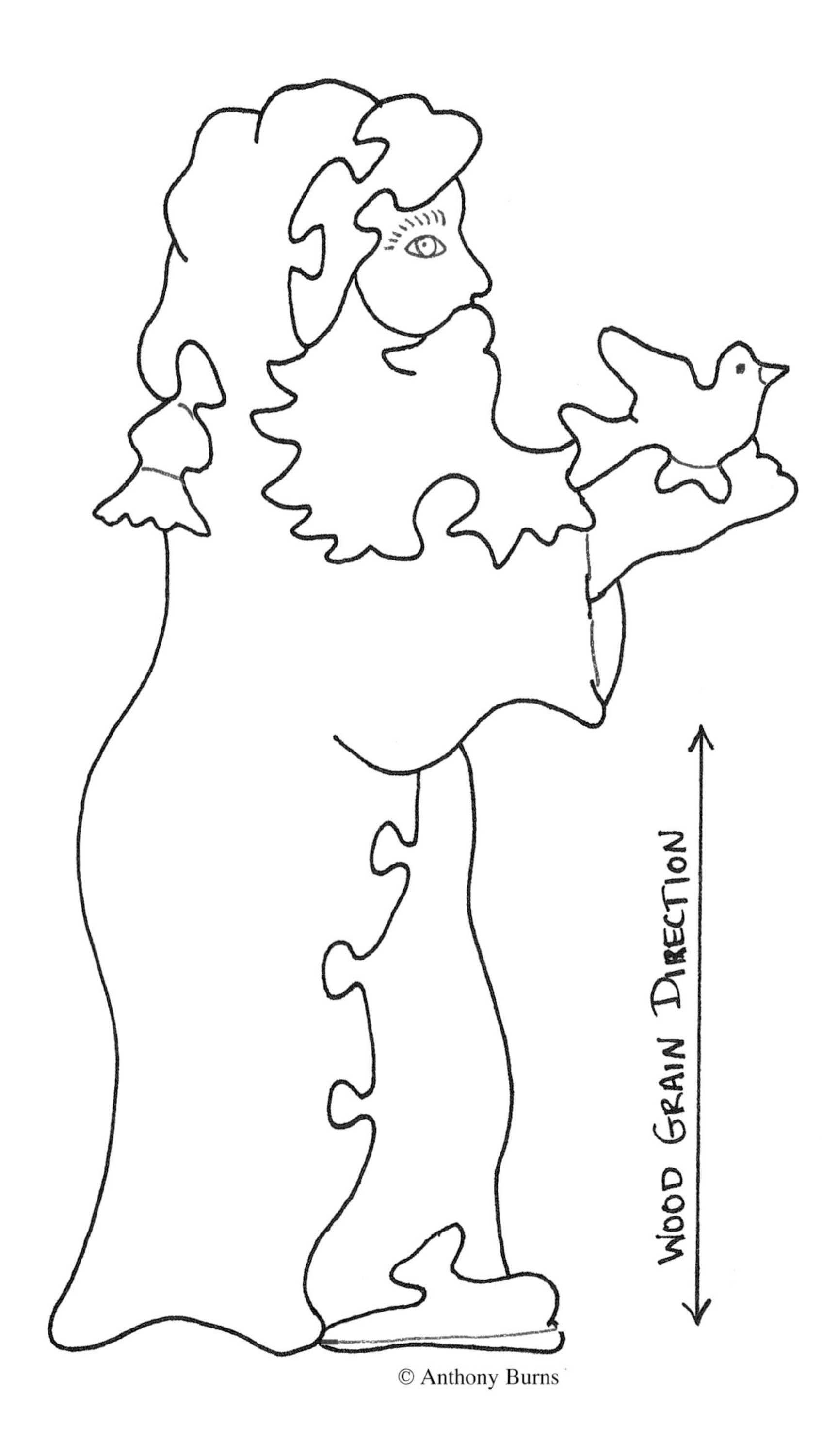

Red line - Detail line
Black line - Cut line

Christmas

Crèche

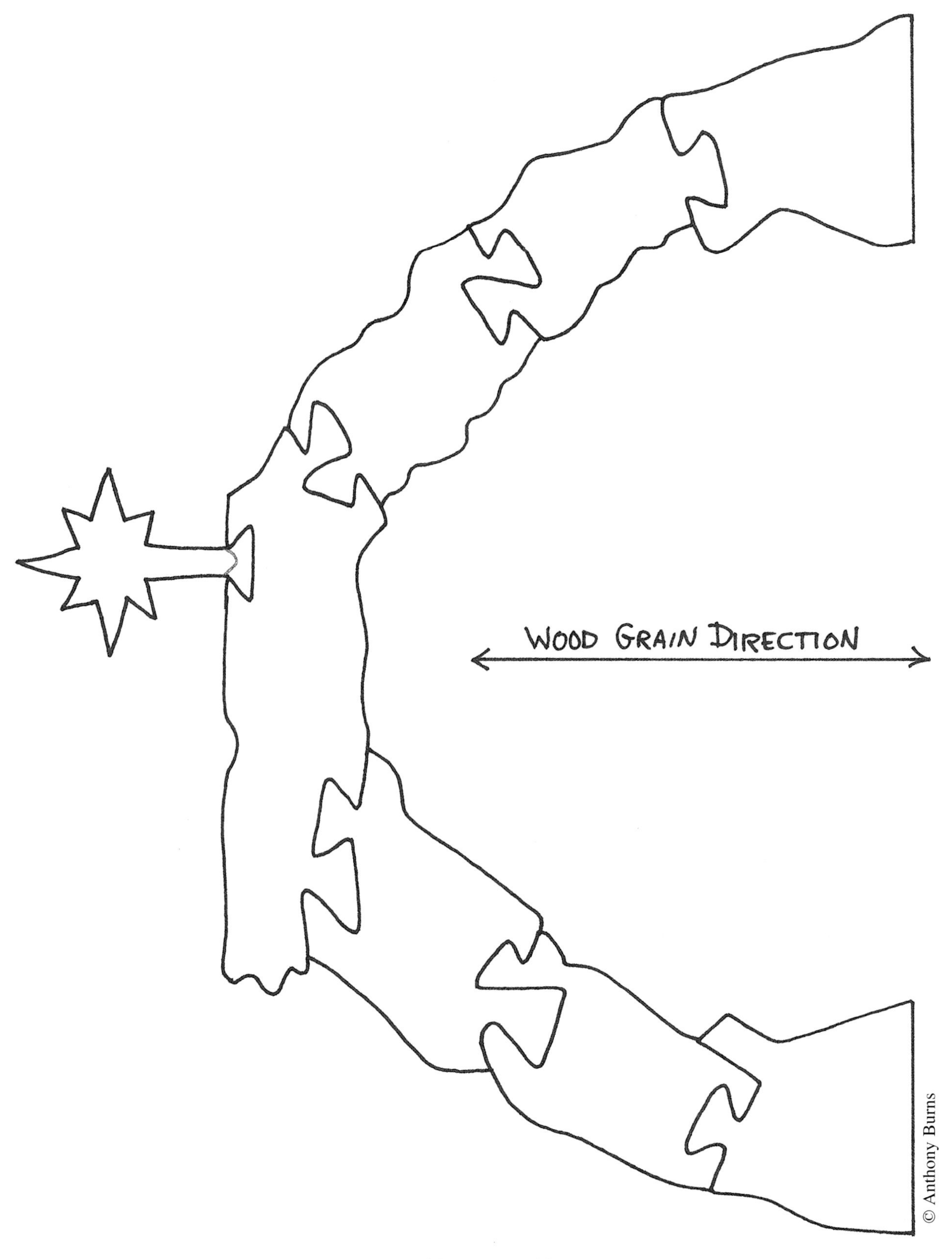

Red line - Detail line
Black line - Cut line

The patterns for the crèche appear on pages 54 to 63.

Joseph

Red line - Detail line
Black line - Cut line

Mary

Red line - Detail line
Black line - Cut line

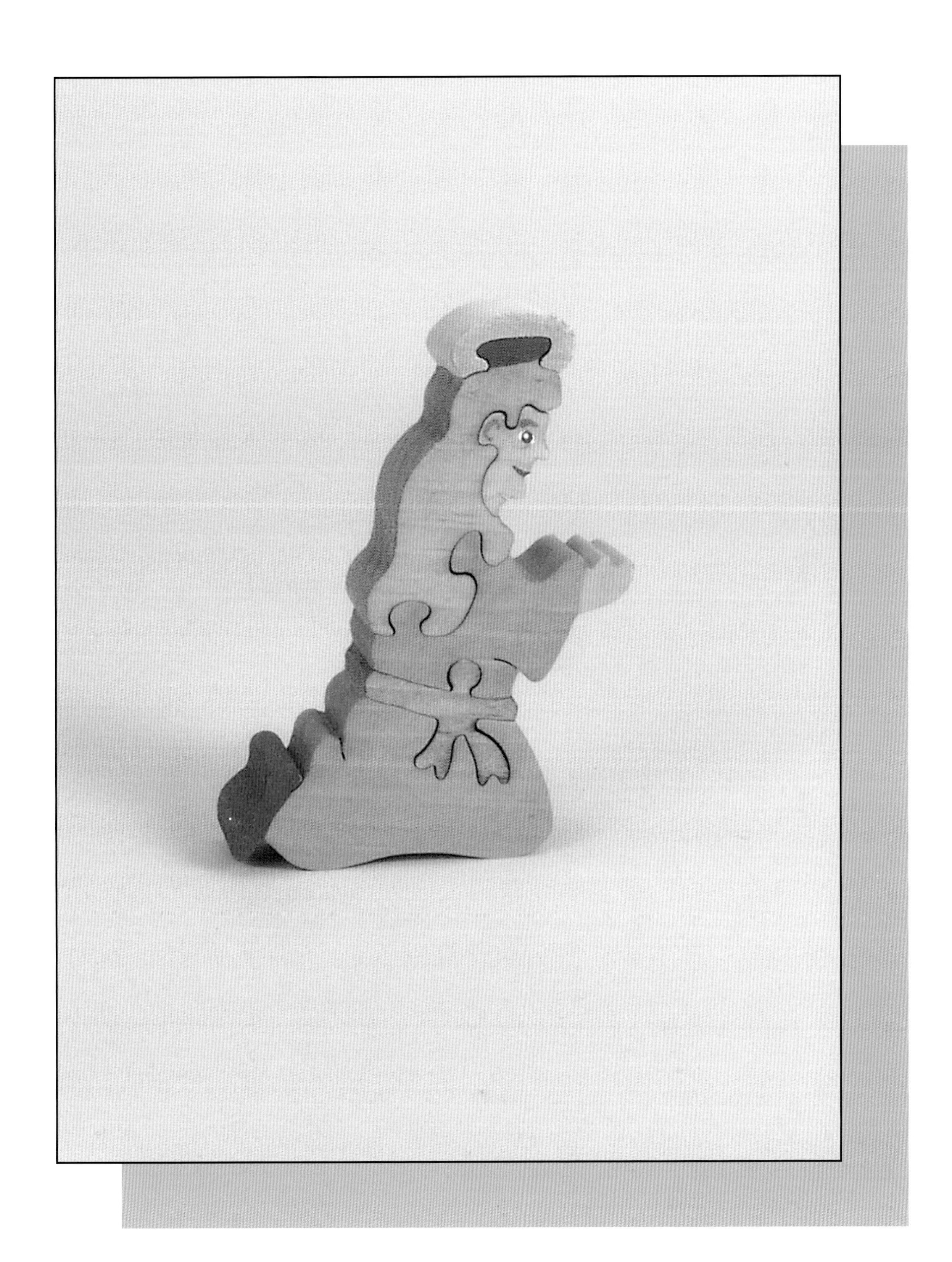

Christmas

Jesus

WOOD GRAIN DIRECTION

Red line - Detail line
Black line - Cut line

Christmas

Angel

WOOD GRAIN DIRECTION

Red line - Detail line
Black line - Cut line

Marriage

Wedding Bells

Red line - Detail line
Black line - Cut line

Birthday Cake

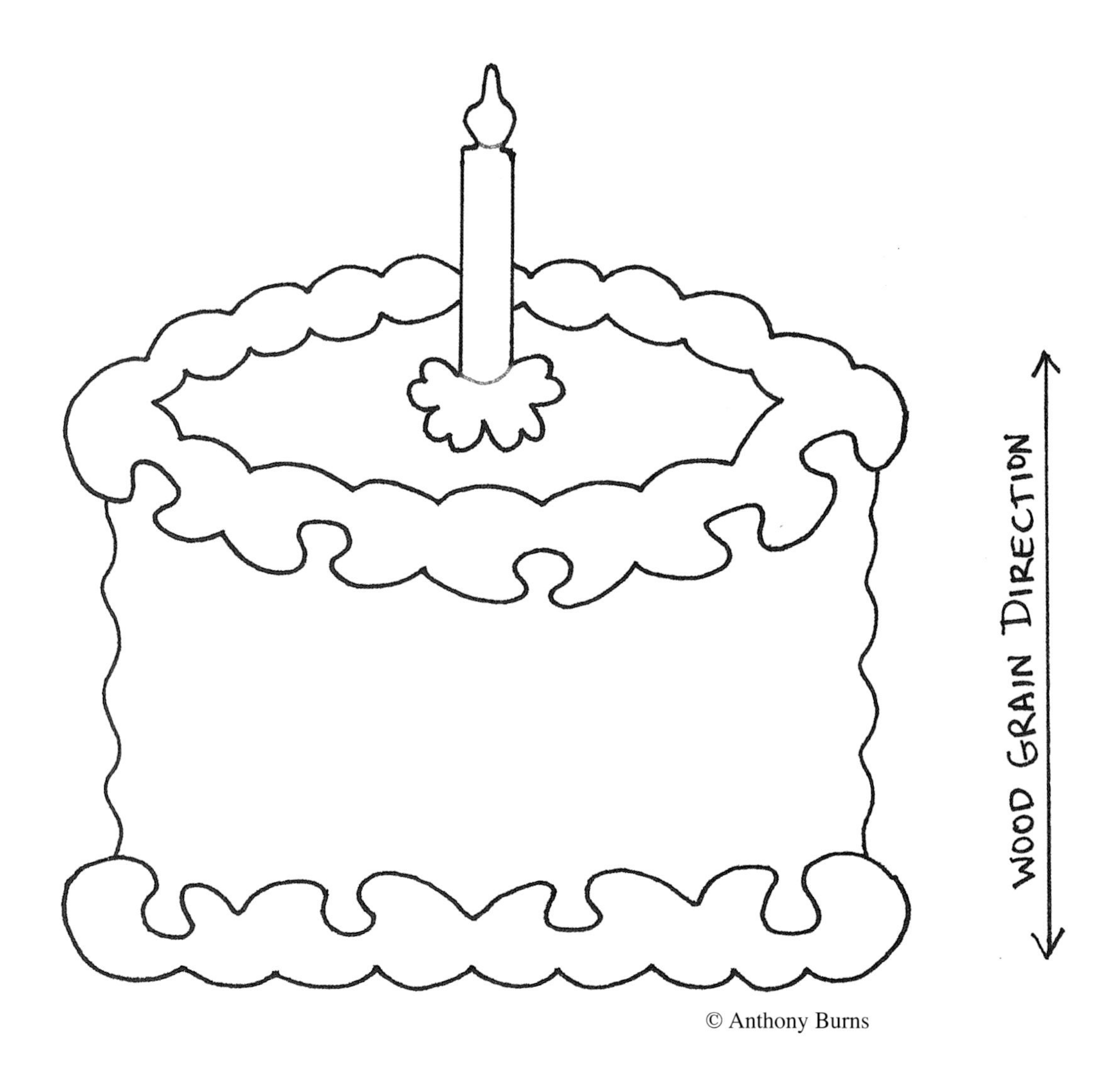

Red line - Detail line
Black line - Cut line